Key Stage 2 Maths – Arithmetic and Algebra

For 10 and 11 year olds with many examples, test questions and answers

Copyright © 2016

E-book editions may also be available for this title. For more information email: valinasser@gmail.com

All rights reserved by the author. No part of this publication can be reproduced, stored in a retrieval system, or transmitted in any form or by any means, electronic, mechanical, photocopying, recording or otherwise, without the prior permission of the publisher and/or author.

ISBN-13: 978-1540893475

ISBN-10: 1540893472

> The author will also do his best to review, revise and update this material periodically as necessary. However, neither the author nor the publisher can accept responsibility for loss or damage resulting from the material in this book

Introduction

This book Key Stage 2 Maths is for 10 and 11 year olds although some of the initial work will also be useful for 9 year olds. It concentrates mainly on Arithmetic and Algebra, these being the building blocks of mathematics. There are tests on each topic and the last test (Test 26) has mixed questions which should give some indication of how much a given pupil has progressed. Aim to get at least 60 -70% of the questions right. Each test is good for a starter lesson or for a 10-15 minute homework or indeed self-study. It starts very gently and gradually builds up to the new curriculum level required in these topics.

About the Author:

The author of this book has experience in both consultancy work and teaching.

Besides working in consultancy he also managed the QTS numeracy tests for teacher trainees at OCR in conjunction with the teaching agency. His book 'Pass the QTS Numeracy Test with ease' is very popular with teacher trainees. The author's initial book 'Speed Mathematics Using the Vedic System' has a significant following and has been translated into Japanese and Chinese as well as German. As a specialist mathematics teacher he has also tutored and taught mathematics and statistics in schools as well as in adult education.

- INTRODUCTION ... 2
- NUMBERS: ... 5
- PLACE VALUE CHART WITH DECIMAL VALUES ... 7
- ADDITION ... 8
- SUBTRACTION ... 10
  - Speed Method of Subtraction ... 10
- NUMBER SEQUENCES ... 13
- FACTORS ... 14
- PLACE VALUE WHEN MULTIPLYING AND DIVIDING BY POWERS OF 10, 100, 1000 ... 15
- MULTIPLES ... 17
- POSITIVE AND NEGATIVE NUMBERS USING THE NUMBER LINE ... 20
- WORD PROBLEMS ... 22
- MULTIPLICATION ... 23
  - Division ... 28
- ROUNDING NUMBERS AND ESTIMATING ... 32
- ESTIMATING CALCULATIONS ... 33
- TIME BASED QUESTIONS ... 37
- FRACTIONS ... 40
  - Simplifying fractions ... 43
- ADDING AND SUBTRACTING FRACTIONS ... 45
  - Division of Fractions ... 46
- PERCENTAGES ... 48

PROPORTIONS AND RATIOS 50
Conversions 51
UNITS, WEIGHT AND CAPACITY 52
QUESTION ON CONVERSIONS 53
BASIC ALGEBRA 55
Simplifying algebraic expressions 56
ALGEBRAIC SUBSTITUTION 57
Simple Equations in algebra 57
ANSWERS 61

Numbers:

Numbers as you recall are made of digits (0, 1, 2, 3, 4, 5, 6, 7, 8 and 9)

Place Value

Thousands	Hundreds	Tens	Units
	2	5	3

Example1: Write the number two hundred and fifty three

This is simply **253**
Hundreds Tens Units

Example 2: Write three thousand, four hundred and sixty five as a number.

Method: This is written as 3465. (The first digit '3' represents three thousand, the second digit '4' represents four hundred, '6' represents six tens and '5' represents five units.

Example 3: Write in words the number 6342

This as you have probably guessed is **six thousand, three hundred and forty two.**

Example 4: Write the numbers 456, 23, 459, 12, 1031, 98, 21 from smallest to biggest.

Method: Starting with the smallest number first the numbers can be written as **12, 21, 23, 98, 456, 459, 1031 (you can see the numbers get bigger as we go along)**

Test 1

(1) Write the number 3045 in words

(2) Write 2300 in words

(3) Write five hundred and two as a number

(4) Write two hundred thousand as a number

(5) Write thirty two thousand, three hundred and fifty as a number

(6) Write the numbers 256, 13, 469, 23, 198, 103, 1021 from smallest to biggest

(7) Write the numbers 34, -5, 2, 0, 40 from biggest to smallest

(8) Write the numbers 23, 23.5, 62, 42, 40.5, 11 from smallest to biggest

(9) Write the number 7567 in words

(10) Write five thousand, three hundred and ten as a number

Place value chart with decimal values

Example1: Below you can see 241.3 written in a place value chart

Thousands	Hundreds	Tens	Units	.	Tenths	Hundredths	Thousandths
	2	4	1		3		

More examples: 0.3 is three tenths, 0.4 is four tenths, 0.01 is one hundredths, 0.001 is one thousandths, 2.3 is two and three tenths.

Test 2

(1)   Write one tenth as a decimal number
(2)   Write three tenths as a decimal number
(3)   Write one hundredths as a decimal number
(4)   Write two hundredths as a decimal number
(5)   Write one thousandth as a decimal number
(6)   Write one and one tenths as a number with a decimal
(7)   Write two and one tenths as a number with a decimal
(8)   Write three hundredths as a decimal number
(9)   Write 4.2 in words
(10) Write 5.01 in words

Addition

I am sure you know how to add but here is another method that is sometimes very useful.

Compensating or adjusting method

In this method we simply adjust by adding or subtracting from the rounded up or rounded down number as shown in the examples below. In example1 we round up 16 to 20 and adjust by taking away 4. Similarly we round up 29 to 30 and adjust by taking away 1. See below for all the working out.

Example 1
18 + 29 =
20 – 2 + 30 – 1 =
20 + 30 – 3 =
50 – 3 = 47

Example 2
19 + 88 =
20 – 1 + 90 – 2 =
20 + 90 – 3 =
110 – 3 = 107

Question on Addition

A teacher buys three maths books A, B and C. The costs are as follows: A costs £3.90, B costs £13.85 and C costs £11.95. Find the total cost of the three books.

Method:

Total cost = £3.90 + £13.85 +£11.95

= £4 - 10p +£14 - 15p + £12 – 5p = £4 +£14 +£12 - 10p - 15p - 5p

= £30 – 30p = £29.70

Test 3 – (You can use long addition or the adjusting method)

(1) 317 + 21 =
(2) 478 + 101 =
(3) 7.3 + 10.2 =
(4) 42 + 117 =
(5) 698 + 3401 =
(6) 3501 + 705 =
(7) 462 + 223 =
(8) 4840 + 445 =
(9) 800 + 16 + 234 =
(10) 4.78 + 1000 =

Subtraction

You probably remember column subtraction and the number line method from your teacher when you last did subtraction. However there is a *'Speed Method'* of doing subtraction which is sometimes useful. **But first let us revisit the familiar method for subtraction.**

Example1:

Work out: 241 - 28

Traditional column method

Consider the following example:

$$\begin{array}{r} 241 \\ -28 \\ \hline 213 \end{array}$$

Starting from the right hand side we cannot subtract 8 from 1 so we borrow 1 from the tens column to make the units column 11. Subtracting 8 from 11 gives us 3. However since we have taken away 1 from the tens column we are left with 3 in this column. Subtracting 2 from 3 in the tens column gives us 1. Since we have nothing else to take away from the hundreds column the final answer is 213.

Speed Method of Subtraction

Example1: Now consider the same problem using a *Speed Method*.

If we add 2 to the top and bottom number we get:

$$\begin{array}{rl} 243 & (241+2) \\ -30 & (28+2) \\ \hline 213 & \end{array}$$

You can see that subtracting 30 from 243 is easier than subtracting 28 from 241!

So essentially we try and add or subtract a certain number to both the numbers in order to make the sum simpler. A few more examples will help.

Example 2:

$$113$$
$$-6$$
$$\overline{}$$

Add 4 to both numbers (**we want to try to make the units column 0 in the bottom row if we can and if it helps**) so the new sum is:

$$117$$
$$-10$$
$$\overline{}$$
$$107$$
$$\overline{}$$

We can see that if we subtract 10 from 117 we get 107.

Example 3:

$$321$$
$$\underline{-114}$$

Let us add 6 to each number so that the unit column in the bottom number becomes a 0 as shown below:

327 (add 6 to 321)

$\underline{-120}$ (add 6 to 114)

207

Subtracting 120 from 327 we get 207 as shown. No borrowing is required.

Note: Sometimes you might find the method above useful; at other times it is easier to revert to the traditional method.

Test 4 – (Use which ever method you prefer)

(1) 232 – 119
(2) 5432 – 789
(3) 7.89 – 6.38
(4) 645 – 457
(5) 678 – 345
(6) £237 - £226.50
(7) £226.65 - £212.50
(8) 5678 – 4597
(9) 6789 – 1237
(10) 7565 – 6557

Number Sequences

Try and see if you can spot the difference between the numbers in a sequence.

Example 1: 11, 14, 17, 20, 23 ___ (the difference between the numbers is +3) so the next number is 26

Example 2: 34, 30, 26, 22, ____ (the difference between the numbers is – 4) so the next number is 18.

Test 5

Complete the following number patterns:

(1) 3, 7, 11, 15, 19 ___
(2) 16, 20, 24, 28, 32, ___
(3) 13, 10, 7, 4, ____
(4) 16, 11, 6, 1, ____
(5) 27, 32, 37, 42, ____
(6) 54, 45, 36, 27, ____
(7) 15, 22, 29, 36, 43, ____
(8) 72, 60, 48, 36, 24, _____
(9) 17, 117, 217, 317, 417, ____
(10) 81.1, 81.2, 81.3, ____ , 81.5

Factors

Factors are simply numbers that will divide exactly into other numbers.

Example 1: Find the factors of 8.

Answer: Factors of 8 = {1, 2, 4, 8}

Example 2: Find the factors of 18.

Answer: Factors of 18 = {1, 2, 3, 6, 9, 18}

Prime Numbers: This is a number with only two factors 1 and itself.

Examples of prime numbers: 3, 5, 7, 11, 13, 17, 23 ….since no other number divides exactly into any one of these.

Test 6

Write down the factors of the following numbers:

(1)   4 =
(2)   6 =
(3)   21 =
(4)   32 =
(5)   70 =
(6)   64 =
(7)   38 =
(8)   15 =

(9)　27 =

(10)　Is 9 a prime number?

Place Value when multiplying and dividing by powers of 10, 100, 1000.

Multiplication

To multiply a number by 10, move the digits one place to the left, to multiply by 100 move all the digits two places to the left. Similarly, to multiply by a 1000, move all the digits three places to the left.

Example 1: 34.6 ×10 = 346

Example 2: 569 ×100 = 56900

Example 3: 67.89 ×1000 = 67890

Division by 10, 100, 1000

To divide a number by 10, move the digits one place to the right, to divide by 100 move all the digits two places to the right. Similarly, to divide by a 1000, move all the digits three places to the right.

Test 7

(1)　677 × 100 =

(2)　56.7 × 100 =

(3)　42 × 100 =

(4)　7.65 × 1000 =

(5)　1000 × 1000 =

(6) 7.89 × 100 =
(7) 76545.345 × 1000 =
(8) 12.567 × 10 =
(9) 8.65 × 1000 =
(10) 965 × 1000 =

Test 8

(1) 456 ÷ 10 =
(2) 17.34 ÷ 10 =
(3) 676 ÷ 1000 =
(4) 441 ÷ 100 =
(5) 456 ÷ 10 =
(6) 1.56 ÷ 10 =
(7) 4.91 ÷ 100 =
(8) 10 ÷ 1000 =
(9) 100 ÷ 1000 =
(10) 8.6 ÷ 1000 =

Multiples

A multiple is a number that may be divided by another a certain number of times without a remainder.

Example 1: Which numbers are multiples of 2 from the numbers shown below?

4, 16, 17, 18.
Clearly 4, 16 and 18 are multiples of 2.

Example 2: Which numbers are multiples of 3 from the numbers shown below?

900, 333, 15, 14, 19

You can probably see that 900, 333 and 15 are multiples of 3

Example 3: Which numbers are multiples of both 2 and 3 from the numbers shown below?

1, 11, 12, 15, 24. **In this case both 12 and 24 are multiples of both 2 and 3**. (Note 15 is only a multiple of 3)

Test 9

You are given the following numbers: 12, 18, 22, 1, 0, 35, 36, 1000, 3000, 2100

(1) Which numbers are multiples of 2?
(2) Which numbers are multiples of 3?
(3) Which numbers are multiples of 11?
(4) Which numbers are multiples of 7?
(5) Which numbers are multiples of both 5 and 7?
(6) Which numbers are multiples of 2 and 3?
(7) Which numbers are multiples of 10?
(8) Which numbers are multiples of 6?
(9) Which numbers are multiples of 9?
(10) Which numbers are multiples of both 2 and 9?

Prime numbers: Let us look at Prime numbers again in more detail.

Prime numbers: is a number that can be divided only by itself and by 1 (without a remainder). For example, 11 can be divided only by 1 and by 11. Prime numbers are whole numbers greater than 1. So for example the first 10 prime numbers are: 2, 3, 5, 7, 11, 13, 17, 19, 23 and 29. **Be careful that an odd number is not necessarily a prime number.** For example **9 is not a prime number** as its factors are 1, 3 and 9 and **prime numbers should have only two factors, 1 and the number itself.** Also, note that 2 is a prime number, the only even number that can be divided by 1 and itself!

Test 10

Which Numbers below are prime numbers?

(1) 5
(2) 7
(3) 9
(4) 12
(5) 2
(6) 17
(7) 21
(8) 31
(9) 21
(10) 32

Positive and Negative Numbers using the number line

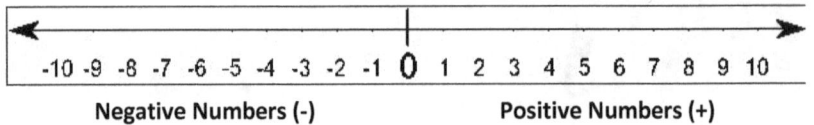

Negative Numbers (-)   **Positive Numbers (+)**

Numbers on the left decrease in value and numbers on the right increase in value.

Examples:

 (1) **3** is smaller than **5**
 (2) **-5** is smaller than **0**
 (3) **-9** is smaller than **–4**

Numbers on the right are bigger than numbers on the left.

Examples:

 (4) **9** is larger than **4**
 (5) **1** is larger than **–1**
 (6) **5** is larger than **–8**

You don't need to always use the number line. You can also use the method below:

When adding and subtracting positive and negative numbers it is worth knowing the following:

When you add two minus numbers you get a bigger minus number.

Example 1: $-4 - 6 = -10$

When you add a plus number and a minus number you get the sign corresponding to the bigger number as shown below:

Example 2: $+6 - 9 = -3$, whereas, $-6 + 9 = 3$

You probably already know the equal sign but there are others too!

= equal sign

> Greater than

< Less than

Examples: (a) 5 > 4, (b) -3 > -5, (c) 8 =8 (d) 0 < 7 ((e) – 17 < - 6

<u>Test 11</u>

Which statements are true?

(1) 8 > 9
(2) 8 < 6
(3) 7 > -1
(4) 7 < -1
(5) 218 < 231
(6) 1 – 7 = -6
(7) -4 -3 = +7
(8) -6 + 3 = -3
(9) +12 – 13 = -2
(10) +12 – 14 = -2

Word Problems

Test 12

(1) I can put 12 marbles in a bag. How many marbles can I put in 7 similar bags?

(2) A teacher can put 15 text books in a box. How many text books can a teacher put 14 boxes of the same size?

(3) John earns £11 per hour doing general maintenance work. The amount of work fluctuates. One good week he works for 43 hours. How much does he earn in this particular week?

(4) What is 300 divided by 6?

(5) How many sevens are there in 147?

(6) John has 400 kilograms of cement which he wants to divide amongst 10 workers. How much does each worker get?

(7) How many elevens are there in 121?

(8) Elizabeth buys some flowers. Each bunch of mixed flowers costs £3.50. Elizabeth buys 6 bunches of mixed flowers. How much does she pay in total?

(9) The maximum temperature in London one December was 13 degrees centigrade and the minimum temperature was -3 degrees centigrade. What was the difference between the highest and lowest temperature?

(10) Arrange the numbers -4, -5, 0, -9, 1, 8, -6 from smallest to biggest.

Multiplication

The Grid Method of Multiplication

This is a very powerful method for those who find traditional long multiplication methods difficult.

Example 1: Multiply 37 × 6

Re-write the number 37 as 30 and 7 and re-write as shown in the grid table.

```
×  | 30    7
---+---------
6  | 180   42
```

Now simply add up all the numbers inside the grid. So the answer is 180 + 42 =222

Example 2: Work out 15 × 13

To work this out using the grid method, re-write 15 as 10 and 5, and 13 as 10 and 3 as shown on the outside of the grid table.

```
×   | 10    5
----+----------
10  | 100   50
3   | 30    15
```

Multiply out the outside horizontal numbers with the outside vertical numbers to get the numbers inside as shown. Finally, just add up the inside numbers which in this case is 100+50+30+15 =195

Multiplication with decimals

Example 3: Work out 1.5 × 1.3

Step1: Leave out the decimal points and just work out the answer to 15 × 13 as shown above.

We know the answer to this is 195.

Step 2: In the answer 195 count two from the left hand side and insert the decimal point.

So the answer is 1.95

If you want to you can think of getting the answer another way:

We know the answer is 195. Note the fact that 1.5 is 15 divided by 10 and 1.3 is 13 divided by 10. So the answer is simply 195 divided by 10×10 =100, so we divide 195 by 100 to get the answer as 1.95

More Multiplication

We will look at some fascinating ways of quickly multiplying by 11, 9, and 5, which will help you speed up your number work in mental arithmetic

Multiplying quickly by 11

One common method used is to multiply by 10 and then add the number itself. We will now look at a super- efficient method that is rarely used.

Super-efficient Speed Method:

11 × 11 =121 **(the first and last digits remain the same & the middle number is the sum of the first two digits)**

The basic method is: Start with the first digit, add the next two, until the last one. This method works with any number of digits.

Let us explore a few more examples with two digit numbers.

13 × 11= 143 (Keep the first and last digit of the number 13 the same, add 1 & 3 to give the middle number 4)

14 × 11= 154

19 × 11= 1(10)9=209 (Notice the middle number is 10, since 1+9=10, so we need to carry 1 to the left hand number.)

A few more examples will show the power of this method.

27 × 11= 297 (the first number=2, the middle number=2+7, the last number =7)

28 × 11=2(10)8= 308 (using similar analysis to 19 × 11 above)

The same principle applies to numbers with more than 2 digits.

Example: Work out 215 × 11

Method: Keep the first and the last digit the same. Starting from the first digit add the subsequent digit to get the next digit, do this again with the second digit until the last digit which stays the same. So, 215 × 11 =2365 (2, is the first digit so stays the same, the sum of 2 and 1 gives you the next digit 3, the sum of 1 and 5 gives you the third digit 6 and finally the last digit 5 stays the same)

Example involving multiplying by 11

There are 54 pupils in year 7 who collect 11 merits each in one month.

How many merits in total did the 54 pupils receive?

54 × 11 using the method explained above is 594

Method: Keep the first and last digit of the number 54 he same, add 5 & 4 to give the middle number 9

Multiplying quickly by 9

Here is an easy method to work out the 9× table

Example 1: Work out 9 × 7

Method:

Step1: Add '0' to the number you are going to multiply by 9, e.g. 7 to get 70

Step2: Now subtract 7 from 70 to get 63 which is the final answer

Example 2: Work out 9 × 12

Method:

Step1: Add '0' to the number you are going to multiply by 9, i.e. 12 to get 120,

Step2: Now subtract 12 from 120 to get 108 which is the final answer

Example 3: Work out 9 × 35

Method:

Step1: Add '0' to the number you are going to multiply by 9, e.g. 35 to get 350

Step2: Now subtract 35 from 350 to get 315 which is the final answer

Essential Method: 6 X 9 = 6(10 −1) = 60 − 6 = 54

A quick way of multiplying by 5

Multiply the number by 10 and halve the answer.

Example 1: 5 × 4 = half of 10 × 4 = half of 40 = 20

Example 2: 5 × 16 = half of 10 × 16 = half of 160 =80

Example 3: 5 × 23 = half of 10 × 23 = half of 230 = 115

TIP: Remember the Order of Arithmetical Operations

Remembering the order in which you do arithmetical operations is very important.

The rule taught traditionally is that of **BIDMAS**.

The **BIDMAS** rule is as follows:

(1) Always work out the **B**racket(s) first
(2) Then work out the **I**ndices of a number (this means squares, cubes, square roots and so on)
(3) Now **M**ultiply and **D**ivide
(4) Finally do the **A**ddition and **S**ubtraction.

Example 1: Work out 2 + 8×3

Do the multiplication before the addition

So 8×3 =24 then add 2 to get 26

Example 2: 4 + 13(7 – 2) this means add 4 to 13×(7 – 2)

Do the **brackets first** so 7 – 2 =5, **then multiply** 5 by 13 to get 65 and **finally add** 4 to get 69

Example 3: work out $3^2 \times 5 - 9$

(Note: 3^2 **means 3 squared or 3 × 3**)

Work out the **square of 3 first**, then **multiply by 5** and finally **subtract 9** from the result.

So we have 3×3 = 9, 9×5 = 45 and finally 45 - 9 = 36

> **Summary**
>
> When working out sums involving mixed operations (e.g. +, -, x and ÷) you need to work out the steps in stages using the BIDMAS rule
>
> So to work out 8 +25 ×12
>
> Do the multiplication first, 25×12 =300, write down 300 then add 8 to get the answer 308.

Division

In general the traditional short or long division approach is a good method. However, there are some other smart techniques worth considering for special situations.

Dividing a number by 2 is a very useful skill, since if you can divide by 2, you can by halving it again divide by 4 and halving it again divide by 8.

Dividing by 2, 4 and 8

Simply halve the number to divide by 2

(Some find it difficult to halve a number like 13. An alternative strategy is to multiply the number by 5 and divide by 10)

Halving again is the same as dividing by 4

And halving once more is the same as dividing by 8

Example 1: 28 ÷ 2 =14

Example 2: 268 ÷ 4 =134 ÷ 2=67

Example 3: 568 ÷ 8=284 ÷ 4=142 ÷ 2=71

Example 4: 65 ÷ 4 =32.5 ÷ 2=16.25

Dividing by 5

An easy way to do this is to multiply the number by 2 and divide by 10.

Example 1: $120 \div 5 = (120 \times 2) \div 10 = 240 \div 10 = 24$

Example 2: $127 \div 5 = (127 \times 2) \div 10 = 254 \div 10 = 25.4$

Similarly to divide by 50 simply multiply by 2 and divide by 100

Dividing by other numbers: The conventional short division method is a good method but you might find the speed methods below useful sometimes.

> **Example:** £67.5 is divided amongst three pupils. How much does each pupil get?
>
> Clearly this is the same as $60 \div 3$ added to $7.5 \div 3$
>
> $60 \div 3 = 20$ and $7.5 \div 3 = 2.5$ which altogether is 22.5
>
> Hence, £67.5 $\div$ 3 = £22.50 per pupil

Example 1:

Divide 145 by 7

(145 = 140 +5)

We can say that $140 \div 7 = 20$, and then we are left with 5/7.

So the answer is 20 and 5/7

Example 2:

Divide 103 ÷ 9

(103 = 99 + 4)

= $99 \div 9 + 4/9$

= 11 and 4/9

Test 13

Work out the following multiplications and divisions without using a calculator. You can use any method you like.

(1) 15 × 27 =

(2) 1.5 × 2.7 =

(3) 45 × 57 =

(4) 123 × 12 =

(5) 32 × 9 =

(6) 1245 × 11 =

(7) 1263 × 2 =

(8) 1111 × 11 =

(9) 35 × 9 =

(10) 140 × 8 =

Test 14

Work out the following

(1) 16 ÷ 4 =
(2) 36 ÷ 9 =
(3) 116 ÷ 2 =
(4) 408 ÷ 4 =
(5) 160 ÷ 4 =
(6) 13 ÷ 2 =
(7) 456 ÷ 4 =
(8) 56 ÷ 8 =
(9) 189 ÷ 3 =
(10) 170 ÷ 4 =

Rounding numbers and estimating

We will start simply with rounding numbers to the nearest 10 and 100

Consider the number 271

Rounded to the nearest 10 this number is 270

Rounded to the nearest 100 this number is 300

(The principle is that if the right hand digit is lower than 5 you drop this number and replace it by 0. Conversely if the number is 5 or more drop that digit and add 1 to the left)

Try a few more:

5382 to the nearest 10 is 5380

5382 to the nearest hundred is 5400

5382 to the nearest 1000 is 5000

This rule can also be applied to decimal numbers:

3.7653 rounded to the nearest thousandth is 3.765

3.7653 rounded to the nearest hundredth is 3.77

3.7653 rounded to the nearest tenth is 3.8

3.7653 rounded to the nearest unit is 4

Tip: remember to use common sense when rounding in real life situations:

Example: A teacher wants to keep 120 English text books in the same size boxes. She can fit 22 text books in a box. How many boxes will she need? You may use a calculator.

Method: Number of boxes required will be 120÷22= 5.5 (to one decimal place). But clearly, she cannot have 5.5 boxes. So she needs to have 6 boxes

Estimating calculations

Example 1: Work out $(2.2 \times 7.12)/4.12$

We can quickly estimate that this is roughly equal to $(2 \times 7)/4$

$=14/4$ which is around 3.5 or 4 rounded to the nearest unit.

The actual answer is: 3.8 (to 1 decimal place)

Example 2: Work out $38 \times 2.9 \times 0.53$

We can approximate 38 to be 40 to the nearest ten

We can approximate 2.9 o 3 to the nearest unit

We can approximate 0.53 to 0.5 to the nearest tenth

So the magnitude of the answer is $40 \times 3 \times 0.5$

This is $120 \times 0.5 = 60$ (approximately)

Test 15

Round these numbers to 2 decimal places.

(1) 3.456

(2) 45.781

(3) 0.0789

(4) 34.985

Round these numbers to the nearest tens

(5) 344

(6) 567

(7) 421

Find the approximate value of these questions:

(8) 2.8 × 5.1

(9) 34.7 × 9.8

(10) (12.89 ×9. 87)÷9.8

Test 16

Write down the squares of the numbers below. (The first one is done for you)

(1) $3^2 = 3 \times 3 = 9$
(2) $4^2 =$
(3) $5^2 =$
(4) $6^2 =$
(5) $7^2 =$
(6) $8^2 =$
(7) $9^2 =$
(8) $10^2 =$
(9) $11^2 =$
(10) $12^2 =$

Test 17

Find the missing numbers:

(1) 20 + ? = 26

(2) ? + 30 = 41

(3) 33 + ? = 50

(4) 78 + ? = 82

(5) 50 + ? = 100

(6) 45 + ? = 45.5

(7) 42 + ? = 80

(8) 45 + ? = 53

(9) 83 + ? = 84

(10) 0 + ? = -1

Time Based Questions

For converting time from 12 hour clock to 24 hour clock see examples below

12 –Hour Clock	24 –Hour Clock
8.45 am	08:45
11.30 am	11:30
12.20pm	12:20
2.35 pm	14: 35 (after 12pm add the appropriate minutes and hours to 12 hours, in this case 2hrs 35mins +12hrs = 14:35)
8.45 pm	20:45 (8hrs 45mins + 12hrs = 20:45)
11.47pm	23:47 (11hrs 47mins +12hrs = 23:47)

The convention is that if the time is in 24-hr clock there is no need to put hrs after the time.

Also remember:

2.5 hours = 2 hours 30minutes (0.5 hours = half of 60 minutes)

2.4 hours = 2 hours 24 minutes (0.4 hours = 0.4×60 = 24 minutes)

For other time based questions e.g. years, months, days, hours, minutes or seconds remember the appropriate units.

Example 1: At a parents evening a teacher has to see the parents of each pupil for 12 minutes. There are 15 pupils. Also there is a break of 20 minutes. The session starts at 5.30pm. When does it finish? Give your answer using the 24 hour clock

Method: Clearly we need to first work out the total time it takes for all the pupils including the break time. Total time for 15 pupils is 15 × 12 = (15×10 +15×2) =180 minutes = 3 hours plus break time of 20 minutes. So the parents evening stops 3hrs and 20 minutes after 5.30pm – this means it ends at 8.50pm. However using the 24 hour clock the times it ends is 20:50

Example 2: At a junior school a child completes a lap in 2.4 minutes. How many minutes and seconds is this?

Convert 0.4 minutes into seconds. Since one whole minute = 60 seconds, then 0.4 minutes = 0.4×60 = 24 seconds. Hence the child completes the lap in 2 minutes and 24 seconds. (Note that 0.4 × 60 is the same as 4 × 6, since if you multiply 0.4 × 10 = 4, correspondingly divide 60 by 10 to get 6)

Test 18

(1) How many hours are there in 180 minutes?
(2) How many seconds are there in 1.5 minutes?
(3) Write 1pm in 24 hour clock
(4) Write 2.30pm in 24 hour clock
(5) Write 9.45pm in 24 hour clock
(6) How many minutes are there in $3\frac{1}{2}$ hours?

(7) A pupil in year 7 completes a lap in 2.5 minutes. How many seconds is this?
(8) How many minutes are there in 300 seconds?
(9) How many seconds are there in 10 minutes?
(10) Write 11.15pm in 24 hour clock

Fractions

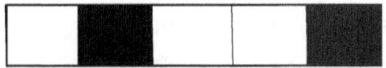

A fraction **is a part of a whole**. So if there are 5 parts altogether and 2 parts are shaded then this can be expressed as 2 out of 5 or $\frac{2}{5}$.

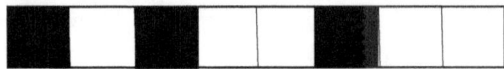

Similarly if 3 parts out of 8 are shaded then this can be written as $\frac{3}{8}$. **The top number is the number of parts you are interested in** and **the bottom number is the total number of parts**.

Example 1: I cut a cake into 9 slices and eat two of them. What fraction of the cake have I eaten?

Method: There are 9 parts in the whole cake. I eat 2 parts out of 9. This means the fraction is $\frac{2}{9}$

I am sure most of you are aware that 'half' can be written as $\frac{1}{2}$

Try and remember the following fractions

$\frac{1}{2}$ (half) $\frac{1}{4}$ (a quarter) $\frac{3}{4}$ (three quarters) $\frac{1}{10}$ (a tenth)

$\frac{1}{3}$ (a third) $\frac{1}{5}$ (a fifth)

Mixed Fractions: This is simply when you combine a whole number with a fraction, for example $2\frac{1}{2}$ (Two and a half) or $1\frac{1}{3}$ (One and a third)

Example 1: I cut an orange into 4 equal parts. I eat one part. Express this as fraction of the whole orange.

Method: There are 4 parts in total so the fraction of the orange that I eat is $\frac{1}{4}$

Example 2: A tutoring group has 7 pupils, four of these are girls.

(1) What fraction of the tutoring group consists of girls?

(2) What fraction consists of boys?

Method: (1) There are 7 people in total in this tutoring group and 4 are girls, this means the fraction that consists of girls is $\frac{4}{7}$

(2) If 4 out of 7 are girls this means 3 out of 7 are boys. So the corresponding fraction of boys is $\frac{3}{7}$

Questions involving fractions

Example 1: Finding a fraction of an amount

(1) Find $\frac{1}{2}$ of £600.

Method: $\frac{1}{2}$ of £600 = $\frac{1}{2}$ × £600 = $\frac{600}{2}$ = £300

Example (2) Find $\frac{1}{5}$ of 400 dollars.

As before to do this, we first write the above question as $\frac{1}{5}$ × 400 = $\frac{400}{5}$ = £80

Test 19

(1) Write down the following fractions, the first two are done for you.
(a) One quarter can be written as $\frac{1}{4}$, (b) Three sevenths can be written as $\frac{3}{7}$, (c) Five elevenths can be written as ……. (d) Three tenths can be written as …….. (e) Four nineteenths can be written as……….
(f) Three fifths can be written as ………….

(2) Find $\frac{1}{2}$ of £600

(3) Lucy cuts a whole cake into twelve parts and 3 people eat seven parts of the cake between them. What fraction of the cake is eaten in total?

(4) Josie is keen to finish her homework before it is due. She manages to complete $\frac{4}{5}$ of it. Write this fraction in words.

(5) Work out $\frac{1}{5}$ of £600

(6) Julie and Andrea receive their weekly pocket money. The total they receive is £5. Julie gets $\frac{3}{5}$ of £5 as she is older than Andrea and remainder is given to Andrea. (a) How much does Julie get? (b) How much does Andrea get?

(7) Write down three quarters as a fraction.

(8) Fiona and Jacob have £35 between them. Fiona has £22. What fraction of the total does Fiona have?

(9) A sweater at Marks & Spencer's is on offer at half price. Its original price was £26. How much will I have to pay if I buy it?

(10) Write down three hundredths as a fraction

Simplifying fractions

Reducing a fraction to its simplest form.

Example 1: Reduce $\frac{20}{50}$ to its form.

Method: Divide top (numerator) and bottom (denominator) by 10 to get $\frac{2}{5}$

Example 2

Reduce $\frac{70}{120}$ to its simplest form. If you divide top and bottom by 10 you get $\frac{7}{12}$

Another example:

Example 3: Simplify $\frac{2}{4}$

Method: You can probably see that $\frac{2}{4}$ is the same as $\frac{1}{2}$. This can be worked out as follows: Divide top and bottom of the fraction $\frac{2}{4}$ by 2. This means the top number becomes '1' and the bottom number becomes '2'. So a simpler way to write $\frac{2}{4}$ is $\frac{1}{2}$

Test 20

Simplify the following fractions:

(1) $\frac{4}{8}$

(2) $\frac{3}{9}$

(3) $\frac{5}{15}$

(4) $\frac{7}{14}$

(5) $\frac{7}{21}$

(6) $\frac{8}{16}$

(7) $\frac{9}{27}$

(8) $\frac{10}{30}$

(9) $\frac{20}{60}$

(10) $\frac{7}{42}$

Adding and Subtracting Fractions

When the bottom numbers (denominators) are the same, just add the top numbers together keeping the bottom number the same. Likewise for subtraction just subtract the top two numbers.

Example 1: $\frac{2}{5} + \frac{1}{5} = \frac{3}{5}$

Example 2: $\frac{2}{5} - \frac{1}{5} = \frac{1}{5}$

When the denominators (bottom numbers) are different we need to find a common denominator.

Example3: Work out $\frac{1}{2} + \frac{2}{5}$

We have to find a number that both 2 and 5 will go into. This is clearly 10.

We can now re-write the fraction with the same common denominator.

To do this we have to ask how did we get the denominator from 2 to 10 for the first part, and likewise for the second part from 5 to 10. The answer is shown below:

$$\frac{1X5}{2X5} + \frac{2X2}{5X2} = \frac{5}{10} + \frac{4}{10} = \frac{9}{10}$$

We had to multiply **top and bottom by 5 for the first part and top and bottom by 2 for the second part** as shown above. We can then add the fraction **as we have the same common denominator.**

Example3: Work out $\frac{3}{7} - \frac{2}{5}$

This is similar to the above except the common denominator is 35. Since both 5 and 7 will go into 35. However this time we have to **subtract**.

$$\frac{3X5}{7X5} + \frac{7X2}{5X7} = \frac{15}{35} + \frac{14}{35} = \frac{29}{35}$$

Multiplying Fractions

Example: $\frac{2}{3} \times \frac{5}{7} = \frac{10}{21}$.

In this case we simply multiply the top two numbers to get the new numerator and multiply the bottom two numbers together to get the new denominator, as shown above.

Division of Fractions

When dividing fractions **we invert the second fraction and multiply** as shown.

Think of an obvious example. If we have to divide ½ by ¼ we intuitively know that the answer is 2. The reason for this is that there are 2 quarters in one half. Let us see how this works in practice.

Example: $\frac{1}{7} \div \frac{1}{2} = \frac{1}{7} \times \frac{2}{1} = \frac{2}{7}$

Test 21

Work out the following fractions:

(1) $\frac{3}{7} + \frac{1}{7}$

(2) $\frac{1}{8} + \frac{6}{8}$

(3) $\frac{1}{5} + \frac{3}{5}$

(4) $\frac{9}{15} + \frac{4}{15}$

(5) $\frac{4}{5} - \frac{3}{5}$

(6) $\frac{1}{2} + \frac{1}{3}$

(7) $\frac{1}{3} + \frac{2}{5}$

(8) $\frac{2}{5} \times \frac{5}{7}$

(9) $\frac{1}{3} \times \frac{2}{9}$

(10) $\frac{1}{2} \div \frac{1}{4}$

Percentages

Percent simply means out of 100. (Per Cent = Per 100)

The symbol used for per cent is %

So 10% means 10 out of 100 or $\frac{10}{100}$

So 12% means 12 out of 100 or $\frac{12}{100}$

To work out percentages can be very useful.

Example 1. Find 10% of £40

We can work this out by writing 10% = $\frac{10}{100}$

So 10% of £40 can be written as $\frac{10}{100}$×40 (Remember 'of' in maths can be written as times 'x'). Simplifying $\frac{10}{100} = \frac{1}{10}$, so $\frac{1}{10}$×40 = £4

Now **Example 2:** Find 50% of £600

Using the method above we get $\frac{50}{100}$×600 = £300 (*Remember:* $\frac{50}{100} = \frac{5}{10} = \frac{1}{2}$) So $\frac{1}{2}$ of £600 is £300

Example 3: I buy a pair of shoes whose original price is £50. However in a sale there is a discount of 20%. How much do I pay for the pair of shoes?

Method: First work out 20% of £50 = $\frac{20}{100}$×50, since 10% of £50 = £5, then 20% of £50 = £10. We now know that the discount is £10. So we pay £50 - £10 = £40 for this particular pair of shoes.

Test 22

(1) Find 20% of £500

(2) Elizabeth buys a coat at 20% discount from the original price of £60. How much does Elizabeth pay for the coat after the discount?

(3) What is 30% of £700?

(4) Calculate 5% of £800

(5) Find 15% of £400

(6) Fatima buys some perfume at 30% discount. The original price is £10. How much does Fatima actually pay after the discount?

(7) What is 75% of 300kg

(8) Find 60% of 200 minutes

(9) John buys a calculator at a discount of 10%. The original price is £4. How much does John pay

(10) Find 90% of £600

Proportions and ratios

Although proportion and ratio are related they are not the same thing – see example below for clarification.

Example: In a class there are 13 girls and 10 boys. The **ratio of girls to boys is** 13:10, and the **proportion of girls in the class** is 13 out of 23 or $\frac{13}{23}$ (Since the total number of pupils is 23, the bottom number is 23)

Questions based on proportions and ratios

Example 1

In a class of 17 pupils, 9 go home for lunch. What is the proportion of pupils in this class that have lunch at school?

Since 9 out of 17 pupils go home, this means 8 pupils have lunch at school.

As a proportion this is 8 out of 17 or $\frac{8}{17}$

Example 2:

£100 is divided in the ratio 1: 4 how much is the bigger part?

The total number of parts that £100 is divided into is 5 (to find the number of parts simply add the numbers in the ratio, which in this case is 1 and 4)

Clearly, 1 part equals £20 (100 divided by 5), so 4 parts is equal to £80 (Since 4×20 =80). So £80 is the bigger part.

Example 3:

£1600 is divided in the ratio of 3 : 5

Find out how much the smallest part is worth?

Clearly £1600 is divided into a total of 8 Parts (add up the ratio parts 3 : 5)

So each part is worth £200 (£1600 divided by 8)

So 3 parts (this is the smallest part) equals 3 ×100 = £300

Example 5:

As we have seen, sometimes ratios are expressed in ways, which may not be the simplest form. Consider 5:10

    (a) You can re-write 5:10 as 1:2 (divide both sides of 5:10 by 5)

    (b) 4: 10 can be re-written as 2: 5 (divide both sides of 4:10 by 2)

    (c) 15 : 36 simplifies to 5 : 12 (divide both sides of 15 : 36 by 3)

Conversions

Conversions are often useful in changing currencies for example from pounds to dollars or euros and vice-versa. It is also useful to convert distances from miles to kilometres or weights from kilograms to pounds and so on.

Basically a conversion involves changing information from one unit of measurement to another. Consider some examples below:

Question based on conversions (You may use a calculator for these sort of questions)

Example 1:

I go to France with £150 and convert this into Euros at 1.2 Euros to a pound.

(1) How many Euros do I get?

Method: (1) Since 1 pound = 1.2 Euros, I get 150 X 1.2 =**180 Euros** in total.

Example 2

The formula for changing kilometres to miles is given by:

$M = \frac{5}{8} \times K$. Use this formula to convert 16 kilometres to miles

Method: substitute K with 16 and multiply by $\frac{5}{8}$

This means $M = \frac{5}{8} \times 16$. This comes to **10 miles**

Units, Weight and Capacity

Metric Measures

1000 Millilitres (ml) =1 Litre(l)

100 Centilitres (cl) =1 Litre (l)

10ml =1 cl

1 Centimetre (cm) =10 Millimetres (mm)

1 Metre (m) = 100 cm

1 Kilometre (km) =1000 m

1 Kilogram (kg) =1000 grams (g)

Imperial Measurements

1 foot =12 inches

1 yard =3 feet

1 pound = 16 ounces

1 stone =14 pounds (lb)

1 gallon = 8 pints

1 inch = 2.54 cm (approximately)

Question on conversions

Example 1:

Method: Since 1000grams = 1kg this means 6000 grams = 6kg

Example 2: How many grams are there in 2.5kg?

Method: Each Kg = 1000g. So 2.5kg = 2.5×1000 g = 2500g

Example 3: How many metres are there in 0.5km?

Method: Since, 1km = 1000m, this means 0.5km = 500m. So there are 500 metres in 0.5km.

Example 4: Robert is going to the USA. He changes £120 into US dollars. The exchange rate is $1.3 to one pound sterling. How many dollars does Robert get?

Method: Since we know that one £ sterling = $1.3 this means £120 = $120×1.3 = $156 (You can use a calculator to check this)

Test 23

(1) Mary invites 28 people to her birthday party. 23 of them are girls. What is (a) the proportion of girls in the party? (b) What is the proportion of boys?

(2) Helen cycles to work 3 days out of 5 during a working week. What is the proportion of days she cycles to work?

(3) Ahmed has sandwiches 5 times a week for his lunch. For 2 days he has curry. What is the proportion of times he has sandwiches for lunch?

(4) £500 is split between John and James in the ratio of 2:3. How much does James get?

(5) £1200 is divided between 2 people in the ratio of 1:2. What is the largest amount that a person gets?

(6) In 2014 in a particular school 79 people take French out of a total of 690 pupils. What is the proportion of pupils who take French in this school?

(7) How many centimetres are there in two metres?

(8) How many metres are there in 1.5 km?

(9) How many metres are there in 2.5Km?

(10) How many grams are there in 7kg?

Basic Algebra

In algebra we often use letters instead of numbers. There are some basic conventions and rules of algebra that you should be familiar with to progress in this subject.

If you see	We Mean
$x = y$	x equals y
$x > y$	x is greater than y
$x < y$	x is less than y
$x \geq y$	x is greater than or equal to y
$x \leq y$	x is less than or equal to y
$x + y$	the sum of x and y
$x - y$	subtract y from x
xy	x times y
x/y	x divided by y
$x \div y$	x divided by y

As we saw in an earlier chapter remember that when adding and subtracting positive and negative numbers it is worth noting the examples below:

When you add two minus numbers you get a bigger minus number.

Example 1: $-4 - 6 = -10$

When you add a plus number and a minus number you get the sign corresponding to the bigger number as shown below:

Example 2: $+6 - 9 = -3$, whereas, $-6 + 9 = 3$

When you subtract a minus from a plus or minus number you need to note the results as shown below:

Example 3: $6 -(-3)$ we get $6+3 = 9$ (since $-(-3) = +3$)

Example 4: $7 -(+3)$ we get $7 - 3 = 4$ (since $-(+3) = -3$)

In this case note that $-(-) = +$. Also, $+(-) = -$ and $-(+) = -$.

Simplifying algebraic expressions

Example 1: Simplify $3x + 4x + 5x$

Method: We simple add up all the x's.

Hence we get $3x + 4x + 5x = 12x$

Example 2: Simplify $3x + 4x + 3y + 5y$

Method: Add up all the like terms.

So we get $3x + 4x + 3y + 5y = 7x + 8y$

(Notice we add up all the x's and then all the y's)

Example 3: Simplify $3m + 4y + 2m - 3y$

Method: as before, we add and subtract like terms.

Now $3m + 2m = 5m$ and $4y - 3y = 1y$ or just y.

So we can write $3m + 4y + 2m - 3y = 5m + y$.

Algebraic Substitution

This is the process of substituting numbers for letters and working out value of the corresponding expression.

Example 1: if a =5 and b=6 work out 2a +3b

Method: Substitute numbers for letters and we get:

2 × 5+3 × 6

(Notice 2a means 2 × a and 3b means 3 × b)

So, 2 × 5 +3 × 6 = 10+18 =28

This means that 2a+3b =28

Example 2: If m=7 and n=8 work out 5m− 3n

Substituting numbers for letters we get:

5 × 7 − 3 × 8 = 35 − 24= 11

So 5m −3n =11

Simple Equations in algebra

Consider the following English statements and their Maths equivalent:

English Statements	**Algebra**
Something plus five equals ten	x + 5 =10
Something times two, plus five equals eleven	2x + 5 = 11

Now consider solving these equations using a common sense approach.

Example 1: Something plus five equals ten. What is 'something'?

Clearly we need to add five to five to get ten. So 'something' in this case equals five.

Solving this by algebra can be very similar. As we saw, we can re-write the English statement above in algebra as follows:

$x + 5 = 10$ (notice, we are representing 'something' by x)

Now, if $x + 5 = 10$ clearly x (which represents 'something') is equal to 5.

So, $x = 5$

Test 24

Find the missing numbers:

(1) $? + 7 = 10$
(2) $7 + x = 11$ (In this case x is the missing number!)
(3) $? - 1 = 22$
(4) $x - 8 = 16$
(5) $x + 11 = 22$
(6) $y + 3 = 15$ (Here y is the missing number)
(7) Solve the equation $x + 6 = 16$ (this means find x)
(8) Solve the equation $x + 10 = 21$
(9) Solve the equation $y + 11 = 15$

(10) Solve the equation x – 2 = 17

Test 25

If a = 5 and b = 8 find the values of:

(1) a + b
(2) 2a
(3) 3b
(4) 2a + 3b
(5) 3b – 2a
(6) 2b – a
(7) 4a + 2b
(8) b – a
(9) 2b – a
(10) 3b + 7a

Test 26 (Final Test) Mixed Questions (No calculators allowed)

(1) Write the number 5345 in words
(2) Work out 14 × 34
(3) What is $\frac{1}{4}$ of £200?
(4) How many grams are there in 8kg?
(5) Find x in the equation x – 9 = 16
(6) Find 20% of £40
(7) Write down two fifteenths as a fraction
(8) If a =7 and b =2 work out a + 2b
(9) Find the next number in the sequence 13, 21, 29, 37, 45, _____
(10) If there are 5 girls and 6 boys in a maths tutoring class what is the proportion of girls?
(11) Work out 515÷5
(12) If £300 is divided in the ratio 1 : 2 find the biggest sum
(13) Write 4.30pm in 24 hour clock form
(14) Write in words 5 > 4
(15) Work out 234.89×100
(16) Work out 309÷100
(17) What is 72÷8?
(18) Write the fraction $\frac{4}{20}$ in its simplest form.
(19) Find the missing number in the equation 17 + ? = 29
(20) Is the number 69 a prime number? Explain your reason.

Answers

Test 1

(1) Three thousand and forty five
(2) Two thousand and three hundred
(3) 502
(4) 200,000
(5) 32,350
(6) 13, 23, 103, 198, 256, 469, 1021
(7) 40, 34, 2, 0, -5
(8) 11, 23, 23.5, 40.5, 42, 62
(9) Seven thousand, five hundred and sixty seven
(10) 5310

Test 2

(1) 0.1
(2) 0.3
(3) 0.01
(4) 0.02
(5) 0.001
(6) 1.1
(7) 2.1
(8) 0.03
(9) Four and two tenths
(10) Five and one hundredths

Test 3

(1) 338
(2) 579
(3) 17.5
(4) 159
(5) 4099

(6) 4206
(7) 685
(8) 5285
(9) 1050
(10) 1004.78

Test 4

(1) 113
(2) 4643
(3) 1.51
(4) 188
(5) 333
(6) £10.50
(7) £14.15
(8) 1081
(9) 5552
(10) 1008

Test 5

(1) 23
(2) 36
(3) 1
(4) -4
(5) 47
(6) 18
(7) 50
(8) 12
(9) 517
(10) 81.4

Test 6

(1) 4 = {1, 2, 4}
(2) 6 = {1, 2, 3, 6}
(3) 21 = {1, 3, 7, 21}
(4) 32 = {1, 2, 4, 8, 16, 32}
(5) 70 = {1, 5, 7, 10, 14, 35, 70}
(6) 64 = {1, 2, 4, 8, 16, 32, 64}
(7) 38 = {1, 2, 19, 38}
(8) 15 = {1, 3, 5,15}
(9) 27 = {1, 3, 9, 27}
(10) 9 is not a prime number as it has more than 2 factors

Test 7

(1) 67700
(2) 5670
(3) 4200
(4) 7650
(5) 1000000
(6) 789
(7) 76545345
(8) 125.67
(9) 8650
(10) 965000

Test 8

(1) 45.6
(2) 1.734
(3) 0.676
(4) 4.41
(5) 45.6
(6) 0.156
(7) 0.0491

(8) 0.01
(9) 0.1
(10) 0.0086

Test 9

(1) 12, 18, 22, 36, 1000, 2100, 3000
(2) 12, 18, 36, 2100, 3000
(3) 22
(4) 35, 2100
(5) 35, 2100
(6) 12, 18, 36, 2100, 3000
(7) 1000, 3000, 2100
(8) 12, 18, 36, 2100, 3000
(9) 18, 36
(10) 18, 36

Test 10

(1) 5 is prime
(2) 7 is prime

(5) 2 is prime
(6) 17 is prime

(7) 31 is prime

Test 11

(3), (5), (6), (8) & (10) are true

Test 12

(1) 84
(2) 210
(3) £473

(4) 50
(5) 21
(6) 40kg
(7) 11
(8) £21
(9) 16 ° Centigrade
(10) -9, -8, -5, -4, 0, 1, 8

Test 13

(1) 405
(2) 4.05
(3) 2565
(4) 1476
(5) 288
(6) 13695
(7) 2526
(8) 12221
(9) 315
(10) 1120

Test 14

(1) 4
(2) 4
(3) 58
(4) 102
(5) 40
(6) 6.5
(7) 114
(8) 7
(9) 63
(10) 42.5

Test 15

(1) 3.46
(2) 45.78
(3) 0.08
(4) 34.99
(5) 340
(6) 570
(7) 420
(8) 15
(9) 350
(10) 13

Test 16

(1) 9
(2) 16
(3) 25
(4) 36
(5) 49
(6) 64
(7) 81
(8) 100
(9) 121
(10) 144

Test 17

(1) 6
(2) 11
(3) 17
(4) 4

(5) 50
(6) 0.5
(7) 38
(8) 8
(9) 1
(10) -1

Test 18

(1) 3 hours
(2) 90 seconds
(3) 13:00
(4) 14:30
(5) 09:45
(6) 210 minutes
(7) 150 seconds
(8) 5 minutes
(9) 600 seconds
(10) 23:15

Test 19

(1) (c) $\frac{5}{11}$, (d) $\frac{3}{10}$, (e) $\frac{4}{19}$, (f) $\frac{3}{5}$
(2) £300
(3) $\frac{7}{12}$
(4) Four Fifths
(5) £120
(6) (a) £3 (b) £2
(7) $\frac{3}{4}$
(8) $\frac{22}{25}$
(9) £13
(10) $\frac{3}{100}$

Test 20

(1) $\frac{1}{2}$
(2) $\frac{1}{3}$
(3) $\frac{1}{3}$
(4) $\frac{1}{2}$
(5) $\frac{1}{3}$
(6) $\frac{1}{2}$
(7) $\frac{1}{3}$
(8) $\frac{1}{3}$
(9) $\frac{1}{3}$
(10) $\frac{1}{6}$

Test 21

(1) $\frac{4}{7}$
(2) $\frac{7}{8}$
(3) $\frac{4}{5}$
(4) $\frac{13}{15}$
(5) $\frac{1}{5}$
(6) $\frac{5}{6}$
(7) $\frac{11}{15}$
(8) $\frac{10}{35}$
(9) $\frac{2}{27}$
(10) 2

Test 22

(1) £100
(2) £48
(3) £210
(4) £40
(5) £60
(6) £7
(7) 225 Kg
(8) 120 mins
(9) £3.60
(10) £540

Test 23

(1) (a) $\frac{23}{28}$ (b) $\frac{5}{28}$
(2) $\frac{3}{5}$
(3) $\frac{5}{7}$
(4) James gets £300
(5) £800
(6) $\frac{79}{690}$
(7) 200 cms
(8) 1500 m
(9) 2500 m
(10) 7000 gm

Test 24

(1) 3
(2) 4
(3) 23
(4) 24
(5) 11
(6) 12
(7) 10
(8) 11
(9) 4
(10) 19

Test 25

(1) 13
(2) 10
(3) 24
(4) 34
(5) 14
(6) 11
(7) 36
(8) 3
(9) 11
(10) 59

Test 26

(1) Five thousand, three hundred and forty five
(2) 476
(3) £50
(4) 8000 grams
(5) 25
(6) £8
(7) $\frac{2}{15}$
(8) 11
(9) 53
(10) $\frac{5}{11}$
(11) 103
(12) £200
(13) 16 : 30
(14) Five is greater than four
(15) 23489
(16) 3.09
(17) 9
(18) $\frac{4}{20} = \frac{2}{10} = \frac{1}{5}$
(19) 12
(20) 69 is not a prime number as it has more than two factors

www.ingramcontent.com/pod-product-compliance
Lightning Source LLC
Chambersburg PA
CBHW061206180526
45170CB00002B/976